YOUR KNOWLEDGE HAS VALUE

- We will publish your bachelor's and
  master's thesis, essays and papers

- Your own eBook and book -
  sold worldwide in all relevant shops

- Earn money with each sale

Upload your text at www.GRIN.com
and publish for free

Felix Dyrek

Biological Investigation: The Variability of Organisms

Measurement of 160 Lima beans

GRIN Verlag

Bibliografische Information der Deutschen Nationalbibliothek:

Die Deutsche Bibliothek verzeichnet diese Publikation in der Deutschen National-
bibliografie; detaillierte bibliografische Daten sind im Internet über http://dnb.d-
nb.de/ abrufbar.

Imprint:

Copyright © 2008 GRIN Verlag GmbH
Druck und Bindung: Books on Demand GmbH, Norderstedt Germany
ISBN: 978-3-640-73516-7

This book at GRIN:

http://www.grin.com/en/e-book/159583/biological-investigation-the-variability-of-
organisms

Felix Dyrek

Biological Investigation: The Variability of Organisms

Introduction:

This bean topic was done in order to show that a species differ in size and mass from even when it developed in the same environment.

Research Question

Which differences does exist in an organism of the same group ?

Hypothesis

If a large number of Lima beans will be meassured in lengh and mass than there will be a small vary up to no vary at all while some will differ much more from the average bean.

Hypothesis Explanation

An organism of a species and the genus wont look exactly the same as another one in nature. When we clone it than but only than its possible. The organism will always differ,even if the differences are small,from another one.

Compounds

160 Lima beans

Variables

length of beans (dependent)
mass of beans (dependent)

Apparatus

1 electronic scale
1 ruler
pencil & paper to record data

Method

Turn on the electronic scale.
Weigh each of the beans and write down the datas
Measure the length of the beans and note each value

Length / cm	Length (sorted) / cm		Mass / g	Mass (sorted) / g
2.4	1.2		1.91	1.07
2.3	1.5		1.19	1.1
2.1	1.72		1.79	1.13
2.2	1.8		1.76	1.13
2.4	1.8		1.86	1.14
2.3	1.9		1.87	1.16
2.6	1.9		1.52	1.19
2.2	1.9		1.53	1.2
2.2	1.9		1.24	1.21
2.4	1.9		2.09	1.23
2.1	2		1.96	1.23
2.4	2		1.95	1.23
2.2	2		1.68	1.24
2.3	2		1.38	1.24
2.4	2		1.51	1.24
2.5	2		1.94	1.24
2.2	2		2.14	1.25
2.5	2		1.47	1.25
2.2	2		1.57	1.26
2.4	2		1.56	1.27
2.7	2		1.39	1.27
2.2	2		1.55	1.33
2.6	2		1.59	1.35
2.4	2		1.94	1.35
2.3	2		1.57	1.37
2.2	2		1.84	1.38
2.6	2		1.86	1.38
2.2	2.1		2.12	1.39
1.9	2.1		2	1.39

2.3	2.1		1.48	1.4
2	2.1		1.79	1.42
2.1	2.1		1.74	1.43
2.3	2.1		2	1.44
2.4	2.1		1.55	1.45
2.1	2.1		2.17	1.46
2.5	2.1		1.39	1.46
2.2	2.1		1.23	1.46
2.3	2.1		2.16	1.46
2.3	2.1		1.54	1.47
2.3	2.1		2	1.47
2	2.1		1.68	1.48
2.4	2.1		1.65	1.48
2.1	2.1		1.48	1.48
2.4	2.1		1.95	1.49
2	2.1		1.24	1.5
2.4	2.1		1.49	1.5
2.1	2.1		1.7	1.5
2	2.1		1.27	1.5
2.1	2.1		1.25	1.51
2.4	2.1		1.8	1.51
2	2.1		1.25	1.51
2.2	2.2		1.5	1.52
2.1	2.2		1.95	1.52
2	2.2		2.25	1.52
2.4	2.2		1.61	1.53
2.2	2.2		1.24	1.53
2.5	2.2		1.65	1.53
2.9	2.2		1.23	1.54
2.2	2.2		2.04	1.55
2.4	2.2		1.07	1.55

2.2	2.2		1.45	1.55
2.4	2.2		1.6	1.56
2.5	2.2		1.65	1.57
2.7	2.2		1.71	1.57
2	2.2		2.15	1.57
2.3	2.2		1.46	1.57
2.2	2.2		1.53	1.58
2.1	2.2		1.71	1.58
2.2	2.2		2.28	1.59
1.9	2.2		1.6	1.6
2	2.2		1.85	1.6
2.9	2.2		1.98	1.6
1.5	2.2		1.52	1.6
1.2	2.2		1.27	1.6
1.72	2.2		2.32	1.6
2.5	2.2		1.5	1.61
2.1	2.2		1.26	1.62
2.1	2.2		2.15	1.62
2.2	2.2		2.1	1.63
2	2.2		1.99	1.64
2	2.2		2.02	1.64
2.3	2.2		1.87	1.64
2	2.2		1.55	1.65
1.9	2.2		1.46	1.65
2.2	2.2		1.78	1.65
1.8	2.2		1.46	1.68
2	2.2		1.62	1.68
2.4	2.2		1.38	1.7
2.2	2.2		1.51	1.7
2.2	2.2		1.8	1.71
1.9	2.3		1.64	1.71

2.2	2.3		1.4	1.71
2.2	2.3		1.51	1.74
2.1	2.3		1.84	1.74
2.3	2.3		1.6	1.75
2.2	2.3		1.98	1.75
2.2	2.3		1.35	1.76
2	2.3		1.6	1.76
2.1	2.3		2.24	1.76
2	2.3		1.13	1.78
2.1	2.3		1.21	1.78
2.1	2.3		1.7	1.79
2.1	2.3		1.57	1.79
2.4	2.3		1.64	1.8
2.1	2.3		1.8	1.8
2	2.3		2.3	1.8
2.3	2.3		1.35	1.83
2.1	2.3		1.9	1.84
2	2.3		1.71	1.84
2.1	2.3		1.52	1.84
2.4	2.3		1.74	1.85
1.9	2.3		2.04	1.85
2.1	2.3		1.2	1.86
2.5	2.3		1.6	1.86
2.4	2.4		1.86	1.86
1.8	2.4		1.46	1.86
2.3	2.4		2.25	1.87
2.3	2.4		2.29	1.87
2.2	2.4		1.16	1.9
2.3	2.4		1.24	1.9
2.5	2.4		1.75	1.91
2.5	2.4		2.06	1.92

2.2	2.4		1.98	1.92
2	2.4		1.62	1.94
2.2	2.4		1.44	1.94
2.2	2.4		1.84	1.95
2.2	2.4		1.5	1.95
2.1	2.4		1.92	1.95
2.2	2.4		1.47	1.95
2.1	2.4		1.5	1.96
2.2	2.4		1.23	1.98
2.4	2.4		1.9	1.98
2.3	2.4		1.43	1.98
2.6	2.4		2.01	1.99
2.2	2.4		1.92	2
2.3	2.4		1.78	2
2.3	2.4		1.33	2
2.4	2.5		1.75	2.01
2.2	2.5		1.63	2.02
2.2	2.5		1.86	2.04
2.5	2.5		1.42	2.04
2.1	2.5		2.2	2.06
2.3	2.5		1.37	2.09
2.5	2.5		1.64	2.1
2.2	2.5		1.85	2.12
2.3	2.5		1.14	2.14
2.3	2.5		1.53	2.15
2.5	2.5		1.76	2.15
2.2	2.5		1.95	2.16
2.2	2.5		1.6	2.16
2.1	2.5		1.48	2.17
2.5	2.5		2.25	2.2
2.5	2.6		1.76	2.24

2.4	2.6		1.58	2.25
2.4	2.6		1.57	2.25
2.5	2.6		1.83	2.25
2.3	2.7		1.58	2.28
2.2	2.7		1.13	2.29
2.3	2.9		2.16	2.3
2.2	2.9		1.1	2.32

Table 1: The values of the length and mass of each bean, unsorted and sorted.

The Classes for each variable:

Classes of length.

In order to establish classes for the data:

Take the lowest value and the greatest value, 1.2 cm and 2.9 cm
Find the difference of these values.

$$2.9 - 1.2 = 1.7$$

Divide this result by 8 (the amount of classes that are needed)

$$1.7 / 8 = 0.2125$$

Now, with this value, you should add it starting with the lowest value of length, 1.2, and continue until you have reached the highest value, 2.9. You will have 8 total classes.

1st class: 1.2 + 0.2125 = 1.4125, **1.2 – 1.4125**

2nd class: 1.4125 + 0.2125 = 1.625, > **1.4125 – 1.625**

3rd class: 1.625 + 0.2125 = 1.8375, > **1.625 – 1.8375**

4th class: 1.8375 + 0.2125 = 2.05, > **1.8375 – 2.05**

5th class: 2.05 + 0.2125 = 2.2625, > **2.05 – 2.2625**

6th class: 2.2625 + 0.2125 = 2.475, > **2.2625 – 2.475**

7th class: 2.475 + 0.2125 = 2.6875, > **2.475 – 2.6875**

8th class: 2.6875 + 0.2125 = 2.9, > **2.6875 – 2.9**

Number of class	Class	Number of elements in class	Median of values in each class / cm
1	1.2 – 1.4125	1	1.2
2	> 1.4125 – 1.625	1	1.5
3	> 1.625 – 1.8375	3	1.8
4	> 1.8375 – 2.05	22	2
5	> 2.05 – 2.2625	63	2.2
6	> 2.2625 – 2.475	47	2.3
7	> 2.475 – 2.6875	19	2.5
8	> 2.6875 – 2.9	4	2.7

Table 2: The number of elements and median for each class of length.

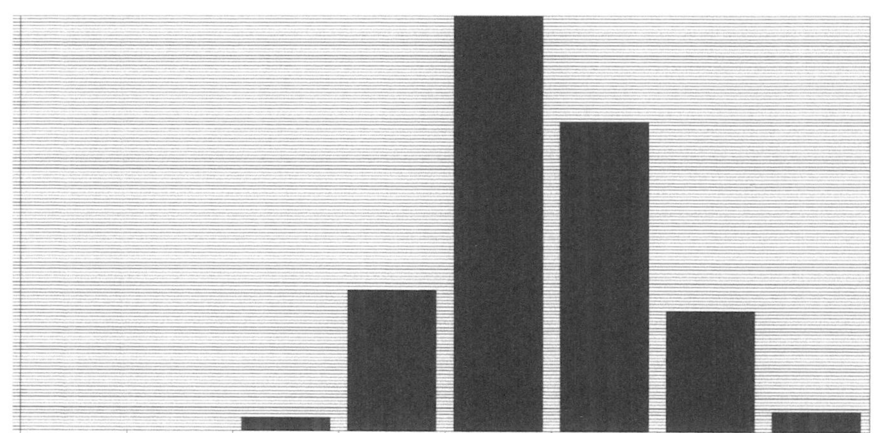

Graph 1: representing the relation between the number of elements and each class of length.

<u>Standard deviation of the values of length:</u> **0.22**

This was calculated by using Excel.

Classes of mass.

Use the same method that was used for classes of length:

Take the lowest value and the greatest value, 1.07g and 2.32g.
Find the difference of these values.

$$2.32 - 1.07 = 1.25$$

Divide this result by 8 (the amount of classes that are needed)

$$1.25 / 8 = 0.15625$$

Now, with this value, you should add it starting with the lowest value of length, 1.07, and continue until you have reached the highest value, 2.32. You will have 8 total classes.

1st class: $1.07 + 0.15625 = 1.22625$, **1.07 – 1.22625**

2nd class: $1.22625 + 0.15625 = 1.3825$, **>1.22625 – 1.3825**

3rd class: $1.3825 + 0.15625 = 1.53875$, **>1.3825 – 1.53875**

4th class: $1.53875 + 0.15625 = 1.695$, **>1.53875 – 1.695**

5th class: $1.695 + 0.15625 = 1.85125$, **>1.695 – 1.85125**

6th class: $1.85125 + 0.15625 = 2.0075$, **>1.85125 – 2.0075**

7th class: $2.0075 + 0.15625 = 2.16375$, **>2.0075 – 2.16375**

8th class: $2.16375 + 0.15625 = 2.32$, **>2.16375 – 2.32**

Number of class	Class	Number of elements in each class	Median of values in each class / g
1	1.07 – 1.22625	9	1.14
2	>1.22625 – 1.3825	18	1.26
3	>1.3825 – 1.53875	30	1.48
4	>1.53875 – 1.695	30	1.6
5	>1.695 – 1.85125	25	1.78
6	>1.85125 – 2.0075	25	1.94
7	>2.0075 – 2.16375	13	2.1
8	>2.16375 – 2.32	10	2.25

Table 3: The number of elements and median for each class of mass.

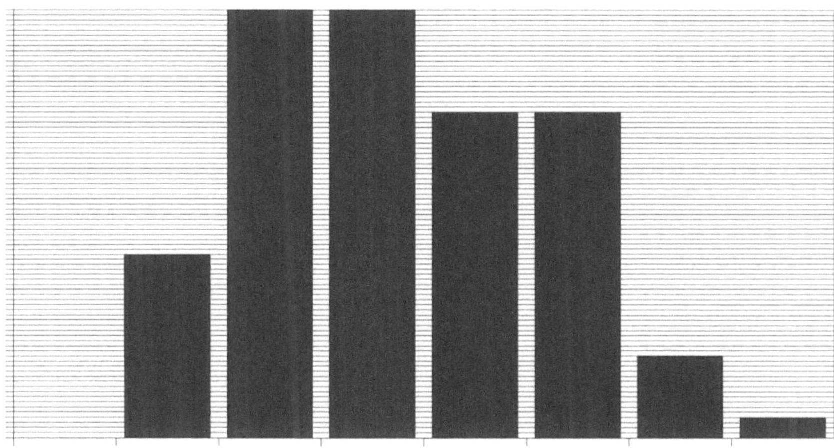

Graph 2: representing the relation between the number of elements and each class of mass.

<u>Standard deviation of the values of mass:</u> **0.3**

This was calculated by using Excel.

Conclusion and Evaluation

My hypothesis that organism of the same species will differ in length and mass but are more or less quite so similar was correct.The measurements prrove it.Only 1.7cm and 1.25gr is the difference between the smallest and largest bean during this experiment.The beans were very close to each other in length and also in mass but still the differences exist.In general there were no obstructions during this experiment.THe only thing which is worth to be mentioned is that we didnt know if these beans which were brought up were the same and we also had no idea about the enviroment the beans were before. The Investigation was succesfull.